EQUATIONS

Maurice F. Stanley

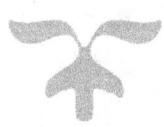

Maurice F. Stanley   Equations

Maurice F. Stanley   Equations

Copyright 2021

Maurice F. Stanley

    ISBN#13:9798580237695

$$W = \int_{k<\Lambda} [Dg][DA][D\psi][D\Phi] \exp\left\{ i\int d^4x\sqrt{-g}\left[\frac{m_p^2}{2}R\right.\right.$$

$$\left.\left. -\frac{1}{4}F_{\mu\nu}^a F^{a\mu\nu} + i\bar{\psi}^i\gamma^\mu D_\mu\psi^i + \left(\bar{\psi}_L^i V_{ij}\Phi\psi_R^j + \text{h.c.}\right) - |D_\mu\Phi|^2 - V(\Phi)\right]\right\}$$

- quantum mechanics
- spacetime gravity
- other forces
- matter
- Higgs

Dedicated

to Glana

and

to my brother

Stephen C. Stanley

and

to the memory of my father and mother

Mr. and Mrs. Claude L. and Louise J. Stanley

Contents

I. Meaning-as-Use 6

II. The Meaning of the Equations of Science and Mathematics 13

III. Wittgenstein's Philosophy of Mathematics 30

III. The Big Bang Loves You 78

IV. A Terrible Sublimity: Kant and the Big Bang 87

I. Meaning-as-Use

I argue here that the real meaning of the equations of mathematics and physics is not reference to numbers or "physical objects", but that these are assertions of will, as all communication is. This insight derives mainly from Wittgenstein, influenced by Schopenhauer, and from the logician Willard Quine While I think this extension of meaning as use actually is implicit in Wittgenstein's philosophy, he had such a contempt for academic metaphysics, especially of the idealist kind, that he drew back from attacking material objects, mass, spacetime, curvature of spacetime, etc.

writhing, groaning, etc.: If his pain were a thing (like a physical object) hidden in his head, it might make sense to doubt whether it was there.

But it doesn't, usually, so pains (and sense data, etc.) are not things Wittgenstein gives a short (but quite inclusive) list of examples of how language really secures meaning *" Review the multiplicity of language-games in the following examples, and others -- Giving orders, and obeying the– Describing the appearance of an object, or giving its measurements– Constructing an object from a description (a drawing)– Reporting an event– Speculating about an event– Etc." (1953, p. 11e, sec. 23)* I would extend this statement to include *all* linguistic activity – words, phrases, statements, whole essays like this one, books, newspaper stories. Furthermore, I want to stress the influence of Schopenhauer, to the effect that *speech* is a personal, bodily act of will: Schopenhauer says this about will: The body is given in two entirely different ways to the subject of knowledge, who becomes an individual through his identity with it. It is given as an idea in intelligent perception, as an object among objects and beholden to the laws of objects. And it is also given in quite a different way, as that which is immediately known to everyone, and is signified by the word will. Every true act of will is also at once and without exception a movement of the body: he cannot really will the act without being at the same time aware that it manifests itself as a movement of his body. The act of will and the movement of the body are not two different things objectively known, which the bond of causality unites; they do not stand in the relation of cause and effect; but they are one and the same, although given in two entirely different ways – immediately, and again in perception for the understanding. The action of the body is nothing but the act of the will objectified, i.e., passed into perception. It will be shown later that this is true of every movement of the body, not merely those which follow upon motives, but also involuntary movements which follow upon mere stimuli, and, indeed, that the whole body is nothing but objectified will, i.e., will become idea. -*The World as Will and Idea,* Arthur Schopenhauer, *Book Two, pp. 32-33.**

---

*Footnote: Schopenhauer influenced Wittgenstein, as is evident in Notebooks 1914-1916, p.37e.

T

The meaning of the assertion "F=ma" is the use it is put to: asserting a prediction of what the experimental measure of the force will be. There need be no reference to *numbers*, or to physical objects, or universals or abstractions, at all, but to real things like bricks, trees, people, dogs, and so forth. Another example: I am talking (to myself) about how many times I've changed my mind today. I am *not* referring to numbers. The meaning of 2+3=5 here is the use *in* the context (language-game, "form of life") of my concern with being wishy-washy. While I think this *extension* is implicit in Wittgenstein's philosophy, he had such a contempt for academic metaphysics, especially of the idealistic kind, that he drew back from this kind of generalization, and was ill-served by the "analytic philosophy" that emerged after his death. Indeed, some philosophers (Anscombe, Grayling) have suggested that he was something of a metaphysician himself – even an idealist Importantly, W. V. O. Quine also hesitated to give up his materialist ontology. In *From a Logical Point of View* he said that our ontological commitment to material objects is only a "myth," but a useful one. (p. 45)

II. The Meaning of the Equations of Science and Mathematics:

The meaning-as-use of '2+3=5'

According to Wittgenstein, pains, minds, thoughts, etc. are not things or mental objects to which we refer when we say such things as "I am in pain," "What time is it?," "I have changed my mind," etc. Such statements do not refer to my mental objects but usually only mean that I need help, or that you should hurry, or that I once believed a certain way but now I do not believe that anymore, I am wavering, and so on "I changed my mind twice before lunch, and then after lunch I changed it three more times. Since 2+3=5, that means I've changed my mind five times so far, today!" Here I am not referring to any numbers, classes of classes of objects, abstractions, or universals. In this context -- there is *always* a context (language-game, form of life) -- I am asserting that 2+3=5, the meaning of which is not reference to objects but an expression of my concern that I am too indecisive. I am complaining to myself that I can't make up my mind. The ontology of numbers is not at issue I took 2 pills before lunch and three more afterwards. I'm not supposed to take more than four a day, so *that's* what I am really talking about – pills (to insist that pills are "physical objects" is a metaphysical ontological assertion). I am not thereby ontologically committed to the existence of "material objects." And when I say or write "**2+3=5**," I am not referring to numbers.

In the same way, the equations of physics are not references to material objects, either. When a scientist or a teacher tells his students "F=ma," the context ('language-game,' 'form of life'-- in a classroom lecture, a conversation among scientists, etc.). *Outside* such a context the equations are only sounds or marks on paper or a chalkboard, he/she are not using language. The teacher is performing an activity of will, asserting something. Its meaning is its use in the language, one of a great multiplicity of such contexts. Some are ordinary, some are extraordinary So I am extending Wittgenstein's doctrine of meaning as use The teacher might go on to say that if the mass of this brick, or tree, or person, is 150 lbs, and the acceleration is 10 ft./sec2, then, since F= ma, the force at which it hits the ground will be 1500 lb-ft/sec2. The meaning is the use the assertion is put to is a prediction. There need be no reference to numbers or ideas or physical objects, but to bricks. The meaning of a statement is its use in the language – words, statements, whole essays like this one, books, news reports and on and on. Newton's Laws of

Motion:

Sir Isaac Newton

Sir Isaac Newton's law of gravitation says:

$$F = \frac{GMm}{r^2}$$

F = force of gravity
G = gravitational constant (6.67×10^{-11})
M = mass of one object
m = mass of other object
r = distance between the two objects

Newton believed in *forces*, which was an ontological commitment. Is Newton referring to material objects in this formula? No. He is asserting that *if* you plug in certain values for the variables you'll get a certain *prediction*. If, on the other hand, you want to insist that an *apple*, say, or a goose, is a *physical, material*

object, you are doing metaphysics/ontology! What does Meaning-as-Use say about the famous equations of modern physics? It means that certain predictions should turn out a certain way. There need be *no commitment* to the existence of *spacetime*, or numbers, or any such thing. To insist that the equation commits you to believing the metaphysical claim that there are gravitational waves, warps in spacetime, wave-particle, etc. is mistaken. *No ontological commitment is here in the equation. The meaning of the equation is its use in the language of the people who assert it.* You use it to explain, make predictions, etc. That is its meaning: how it is used in doing the work of physicists, teachers, etc.

Einstein

Here is a version of Einstein's General Relativity Equation, which is about curvature of space in the vicinity of big stars, etc.

Einstein equations:

$$G_{\mu\nu} = 8\pi\, G\, T_{\mu\nu}$$

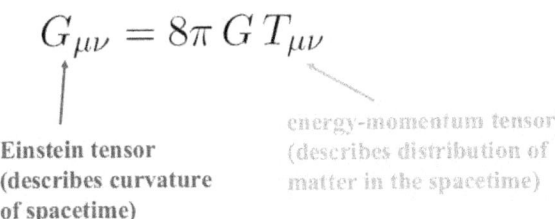

Einstein tensor (describes curvature of spacetime)

energy-momentum tensor (describes distribution of matter in the spacetime)

Schrodinger

Schrodinger's wave equation, about particles, seems to contradict Einstein's Theory of General Relativity equation:

Schrödinger's Equation

$$i\hbar \frac{\partial}{\partial t}\psi(\mathbf{r},t) = -\frac{\hbar^2}{2m}\nabla^2\psi(\mathbf{r},t) + V(\mathbf{r},t)\psi(\mathbf{r},t)$$

i is the imaginary number, $\sqrt{-1}$.
$\hbar$ is Planck's constant divided by 2π: 1.05459×10^{-34} joule-second
$\psi(\mathbf{r},t)$ is the wave function, defined over space and time.
m is the mass of the particle.
∇^2 is the Laplacian operator, $\frac{\partial^2}{\partial x^2} + \frac{\partial^2}{\partial y^2} + \frac{\partial^2}{\partial z^2}$.
$V(\mathbf{r},t)$ is the potential energy influencing the particle.

Maurice F. Stanley   Equations

Hawking:

For metaphysical bias the most blatant offender is (was) Stephen Hawking. One version of his united field theory equation is as follows:

Hawking's final version, done as elegantly as possible goes this way:

$$S = \frac{\pi A k c^3}{2hG}$$

This equation is about physical particles, gravity, spacetime curvature – *everything* physical. But Hawking is talking metaphysics/ontology, not physics! Not that anyone is forbidden to talk metaphysics(!), but Wittgenstein was (understandably) prejudiced against that language-game, which is as permissible as any other, more "ordinary" kind of language. There is an admixture of metaphysics, *ontological* talk in this context.

Einstein wants to refer to curvatures in space-time, to *fields*; and Schrodinger wants to talk about *particles* – actually, *probabilities* Both *fields* and *particles* are ontological posits, which Wittgenstein's Meaning-as-Use theory rejects. Is a *probability* a material object? It is plain, I think, that the two equations are not referential or descriptive of material objects and do not need to involve ontological commitments to objects – wave or particles – but must be translated (a la Quine) in different ways, and do not *contradict* each other, except metaphysically.

Ancient gods, numbers, forces, curvatures of space-time, wave particles, probability densities, black holes …? Do these equations really refer to such "things?" Only experiments, predictions, observations, calculations, really count.

Conclusions:

The *meanings* of these equations, in the context of getting the job of science done, are the activities of will to make sense of the world. This statement perhaps needs explanation. That's the basic intent of these assertions. That's how they are actually used. It is not irrelevant to ask for the broader *motive* of scientific assertion. Einstein said it had to do with the *joy of discovery*, the absorption, rising above one's low, ordinary concerns (as Socrates believed). It takes you up and away. Consciousness – self-consciousness – fades into … the inquiry into the "external world," *out* of subjectivity. Einstein was not naïve about metaphysics

Here is a plain (but Wittgensteinian) argument for why I think materialism, the ontological commitment to physical objects, is wrong: I feed crackers to a friendly goose in the parking lot of a local mall (I think I'm not supposed to). Is the goose a material object? That is a *metaphysical* question. He (the goose) is a GOOSE, the cracker is a cracker. *This interaction of me and the goose does not prove any ontological thesis.* This goose has a family which he cares about, etc. A person who sees him as a physical object made of tinier physical objects, or a cluster of sense data, obeying purely deterministic physical laws, is blinding himself to the non-physical realities of our world (freedom, will, etc.). Size, weight, color, etc. – measurable, observable qualities – are *not* the whole truth about reality.

Science is one thing; materialism is something else. The ontological commitments of materialism are no more scientific than those of idealism. We cannot *observe* 'material objects' (forces, gravitons, wave packets, etc.) any more than you can observe minds or pains or souls. But we *can* observe a change (a wobble, a precession) in the perihelion of Mercury, and then look to see whether our equations have predicted it accurately. That is what equations are all about.

GODEL

"Provable" and "true" go with STATEMENTS, not Numbers!

III. Wittgenstein's Philosophy of Mathematics

(Review of

Michael Potter, "Wittgenstein on Mathematics"

in

The Oxford Wittgenstein Handbook)

There seems to be almost a consensus among mathematicians and philosophers that there is nothing we can confidently call Wittgenstein's philosophy of mathematics. This is strange because Wittgenstein himself wrote volumes on it and believed that his primary contribution to philosophy was in the philosophy of mathematics.

I will argue here that there is indeed a coherent Wittgensteinian philosophy of mathematics, which I will try to present in outline an answer to this mistaken view of Wittgenstein's later philosophy of mathematics. To do this I will have to contradict two splendid essays, one by Professor Michael Potter of Cambridge University and another by Dr. Sorin Bangu of the University of Bergen in Norway.

Potter says:

Although all of Wittgenstein's philosophy no doubt presents difficulties to the interpreter, the secondary literature on his philosophy of mathematics is peculiarly inclusive even when it is judged by these standards. Why is this? – p.122.

The Remarks on the Foundations of Mathematics contain many thought-provoking observations and ideas, and anyone interested

in the philosophy of mathematics ought to read them, but they are

of variable quality and can hardly be said to present something

that amounts to a coherent view. – p. 135.

We know what he was against: Platonism, logicism, intuitionism, and Hilbertian formalism all at various times come in for his criticism. But what, by contrast, was he for? We can be

fairly sure, I think, that he never gave up the two views that were central to his early philosophy if mathematics: that numbers are not objects, and that arithmetical equations are not tautologies. But if it is the first task of any philosopher of mathematics who holds these two views to give an account of arithmetical generalizations, and if Wittgenstein eventually gave up his account according to which

their meanings are given by their proofs, then it is hard to see

what he had to offer instead. Indeed, one struggles to present, even in outline, a positive account of mathematics that can reasonably be called Wittgensteinian. – p. 136.

Potter, Michael, "Wittgenstein on Mathematics," Chapter 6, *The Oxford Handbook of Wittgenstein*, Edited by Oskari Kuusela and Marie McGinn, Oxford University Press, 2011.

Bangu says:

In fact, it is not even clear that the threads of his thinking on mathematics, when pulled together, amount to what we would

today call a coherent, unified-ism. However, one view that can be attributed to him is that mathematical identities such as "Three times three is nine" are not really propositions, as their superficial form indicates, but are certain kinds of rules; and, thus understood, the question is whether they are arbitrary or not. The interpretive position preferred in this article is that they are not, since they are grounded in empirical regularities – hence the recurrence of the theme of the applicability of mathematics in Wittgenstein's later reflections on this topic.

As a first approximation, for Wittgenstein arithmetical identities (such as 'three times three is nine') are not propositions, as their superficial grammar indicates – but rules. Importantly though, these rules are not arbitrary; in a sense (to be explicated later on), the rules in place are the only ones that could have been adopted

or, as Steiner [2009, 12] put it, "the only rules available." The typical conventionalist difficulty (that they might have an arbitrary character) is answered when it is added that the rules are grounded in objectively verifiable empirical regularities (Fogelin [1987]; Steiner [1996], [2000], [2009]; or, as Wittgenstein says, the empirical regularities are "hardened" into rules (RFM VI-22). In discussing Wittgenstein's relation to conventionalism, a central

task in what follows will be to clarify why (and how, and what kind of) agreement within a community is a crucial presupposition of the very existence of mathematics.

..., it is a constant of Wittgenstein's view that mathematics cannot be separated from a language and a human practice.

Bangu, Sorin. *Wittgenstein, Ludwig: "Later Philosophy of Mathematics," Internet Encyclopedia of Philosophy,*

But this is mostly mistaken!

"The meaning of a statement is its means of verification"

– that's the *Tractatus* view, which Wittgenstein abandoned. His final theory is that the meaning of a unit of language is its

practical use in a *context.*

There is *no context* for Gödel's theorems, *no* language game, *except* that of axiomatic logic. Potter thinks mathematical

induction is the *meaning* of a mathematical statement, but this contradicts meaning-as-use, and is mistaken. The assertion

"2+3=5" is a tautology, its denial is unthinkable. It is certain

because of the meanings of the words involved. It is not about *numbers,* because

it does not refer. Its meaning comes from its *use.* Gödel was wrong, confused.

Gödel says: *The development of mathematics in the direction of greater exactness has – as is well known – led to large tracts of it becoming formalized, so that proofs can be carried out according*

to a few mechanical rules. The most comprehensive formal

systems yet set up are, on the one hand, the system of Principia

Mathematica (PM) and, on the other, the axiom system for set theory of Zermelo-Fraenkel (later extended by J. v. Neumann).

These two systems are so extensive that all methods of proof used in mathematics today have been formalized in them, i.e. reduced

to a few axioms and rules of inference. It may therefore be

surmised that these axioms and rules of inference are also

sufficient to decide all mathematical questions which can in any way at all be expressed formally in the systems concerned. It is shown below that this is not the case, and that in both the systems mentioned there are in fact relatively simple problems in the

theory of ordinary whole numbers which cannot be decided from

the axioms.

- *On Formally Undecidable Propositions Of Principia Mathematica And Related Systems,* Kurt Gödel, p. 37-38.

Another expression of Gödel's Theorems:

Gödel's Incompleteness Theorems

1. *First incompleteness theorem*

Any consistent formal system F within which a certain amount of elementary arithmetic can be carried out is incomplete; i.e., there are statements of the language of F which ca neither be proved nor disproved in F.

2. Second incompleteness theorem

For any consistent system F within which a certain amount of elementary arithmetic can be carried out, the consistency of F cannot be proved in F itself.

- Godel's Incompleteness Theorems *(1931) The End of Hilbert's Dream,* Quora Digest Internet 2016.

Wittgenstein says:

It is my task, not to attack Russell's logic from within, but from without.

That is to say: not to attack it mathematically – otherwise I should be doing mathematics– but its position, its office.

My task is, not to talk about (e.g.) Gödel's proof, but to pass it by.

- *Remarks on the Foundations of Mathematics* by Ludwig Wittgenstein, 174.

To show that Potter, Bangu, *et al*, are mistaken, I argue here

that the real meaning of the equations of mathematics and physics

is not reference to numbers ("mental objects") or wave packets ("physical objects"), but that these are assertions of will, as all communication is. This insight derives mainly from Wittgenstein, influenced by Schopenhauer, and to some extent from the logician Willard Quine.

While I think this extension of meaning as use is implicit in Wittgenstein's philosophy, he had such a contempt for academic metaphysics, especially of the idealist kind, that he drew back from attacking material objects, curvatures of spacetime, forces,

etc. Suppose we observe a man writhing, groaning, etc. and he

cries:

"Oh, God, **the PAIN**! It's SO severe! In my CHEST! Oh, GOD,

OH, NO! OH, JESUS, I'm DYING!"

If his pain were a *thing* (like a physical object) hidden in his chest, it might make sense to doubt whether it was there. But it doesn't, usually, so pains (and sense data, etc.) are not things, and the practical use-meaning of that utterance is usually, ordinarily, a *cry* for help.

Wittgenstein gives a short (but quite inclusive) list of examples of how language really secures meaning:

" *Review the multiplicity of language-games in the following examples, and others:*

-- *Giving orders, and obeying them*
– *Describing the appearance of an object, or giving its measurements*

– *Constructing an object from a description (a drawing)*

– *Reporting an event*

– *Speculating about an event*

– *Etc."*

(1953, p. 11e, sec. 23) – Philosophical Investigations

I would extend this statement to include *all* linguistic activity – words, phrases, statements, whole essays like this one, books, newspaper stories, and on and on. Furthermore, I want to stress the influence of Schopenhauer, to the effect that *speech* is a personal, bodily act of will:

Schopenhauer says this about will:

The body is given in two entirely different ways to the subject

of knowledge, who becomes an individual through his identity

with it. It is given as an idea in intelligent perception, as an object among objects and beholden to the laws of objects. And it is also given in quite a different way, as that which is immediately known

to everyone, and is signified by the word will. *Every true act of*

will is also at once and without exception a movement of the body: he cannot really will the act without being at the same time aware that it manifests itself as a movement of his body.

- Arthur Schopenhauer, *The World as Will and Idea,* Book Two, pp. 32-33*.

*Footnote: Schopenhauer's influence on Wittgenstein is evident in Notebooks 1914-1916, p.37e.

The meaning of the assertion "F=ma" is the use it is put to: asserting a prediction of what the experimental measure of the force will be. There need be no reference to *numbers*, or to

"physical objects," or universals or abstractions, at all, but to real things like bricks, trees, people, dogs, and so forth. (To insist that a brick is a

"physical object" is a metaphysical/ontological claim.)

Another example: I am talking (to myself) about how many times I've changed my mind today. I am *not* referring to minds or times

or numbers. The meaning of the words here is their use in the context (language game, "form of life") of my assertion of my concern with being wishy-washy, indecisive.

Importantly, W.V.O. Quine also hesitated to give up his materialist ontology. But he was aware of the situation, and he remarked on it:

"Both kinds of entities [physical objects and gods] enter our conception only as cultural posits. The myth of physical objects is epistemologically superior to most [other myths]."

- W.V. O. Quine, *From a Logical Point of View*, Harvard

University Press, Cambridge, Mass: 1953, p. 44).

Of course, he could not quite forsake physicalism, but still he was aware that it didn't matter as far as getting the job of science done.

According to Wittgenstein, pains, minds, thoughts, etc. are not things or mental objects to which we refer when we say such things as "I am in pain," "What time is it?" "I have changed my mind,"

etc. Such statements do not refer to my mental objects but usually only mean that I need help, or that you should hurry, or that I once believed a certain way but now I do not believe that way anymore,

I am wavering, and so on.

"I changed my mind twice before lunch, and then after lunch I changed it three more times. Since 2+3=5, that means I've changed my mind five times so far, today!" Here I am not referring to any minds, numbers, classes of classes of objects, abstractions, or universals. In this context -- there is *always* a context (language game, form of life) – I am asserting that 2+3=5, the meaning of which is not reference to objects but an expression of my concern that I am too indecisive. I am complaining to myself that I can't make up my mind.

The ontology of numbers is not at issue.

"Do you have a couple of quarters in your pocket?" is not

usually an expression of pure scientific curiosity. It ordinarily

means something like this – that I need two more to go with the

three I have in my pocket to buy a local newspaper. It is not a reference to objects. It is a cry for help (well, a request, perhaps). The two from your pocket alas the three from mine will equal five, just what I need!

In the same way, the equations of physics are not references to material objects, either. When a scientist or a teacher says "F=ma," the context ('language-game,' 'form of life') might be a classroom lecture, a conversation among scientists, etc.). *Outside* such a context the "equations" are only sounds or marks on paper or a chalkboard. Inside that context the teacher is performing an activity of will, asserting something. Its meaning is its use in the language, one of a great multiplicity of such contexts. Some are ordinary,

some are extraordinary.

So, I am extending Wittgenstein's doctrine of meaning as use.

The teacher might go on to say that if the mass of this brick, or tree, or person, is 150 lbs, and the acceleration is 10 ft./ sec^2, then, since F= ma, the force at which it hits the ground will be 1500 lb-ft/sec^2. The meaning is the use the assertion is put to. It is a prediction. There need be no reference to numbers or ideas or physical objects, but only to bricks.

The meaning of a statement is its use in the language – words, statements, whole essays like this one, books, news reports and on and on. This doctrine also applies to mathematical equations, those of arithmetic and algebra and as well as those of the equations of mathematical physics.

Wittgenstein's doctrine is that meaning is not reference to objects, but is an activity of will, a practical use of words, an intention to get something done. This means that pains, thoughts, minds, numbers, etc. are not mental 'things/objects.' and (I argue) neither are there physical objects. This does *not* mean that there are no geese, or planets, or rocks; but the further claim, that there are physical objects, is a metaphysical/ontological move in the context of a philosophical-argumentation language game.

Wittgenstein's theory of meaning-as-use can be captured quite well, I think, in these quotes:

"For a large class of cases – though not for all – in which we employ the word 'meaning' it can be defined thus: the meaning

of a word is its use in the language."

- (1953, Philosophical Investigations, p. 20e, sec. 43).

"Just try – in a real case –to doubt someone else's fear or pain."

(p. 102e 303)

Wittgenstein does not mean that people do not suffer or

tremble fearfully, it means only that to take meaning as reference

to a pain or a sensation is a mistake we should avoid. Suppose I

see someone writhing, groaning, etc.: If his pain were a thing (like a physical object) hidden in his head, it might make sense to doubt whether it was there. But it doesn't, usually, so pains (and sense data, etc.) are not things.

In the same way Wittgenstein treats 'pains,' we can also treat minds, sensations, times, rights, etc. The assertion, 'I have a pain in my hand' is not just a reference to a thing/object, but it can be a verbal substitute for a cry of pain. In an analogous way, 'I am of

two minds about so-and-so' is not about two countable 'mental objects' at all. It might only mean that I have not decided. These

are simple, obvious points about ordinary, everyday language. But this doctrine can be extended to the language of philosophy as well, and, indeed, to all kinds of verbal communication. All

language is an expression of practical will, from newspaper headlines and science texts to Christmas cards, Bible verses, Koran surahs, or the sayings of Buddha. Meaning-as-use applies to them all.

To ask, 'How much time do we have' is not usually a scientific question about the hands of a clock or the spatial relation of the

sun to the surface of the earth, but is more likely an attempt to

hurry someone, or oneself, or to remind them that it's near dinner. (Although it *can* be a question about what a watch says, of course.) In any case it is an assertion of purpose of some kind, a try at getting something done, an activity of will. [See esp.

Wittgenstein, *Philosophical Grammar, pp. 215 ff.* and *The Big Typescript pp. 281e ff. (sec. 81) and p. 300e ff.*]

About the language of rights and political obligation: The assertion, "You have a right to speak freely" seems to amount to a kind of encouragement, a promise to stand by the person addressed. It is not a reference to objects. "You have a right to cheat, rob and kill to get ahead in life" is not a reference to things called *rights*, which one might carry in one's pocket with some coins, or find inscribed on parchment or stone. (It would seem to be an expression of ill will, bad advice, and the person who says it would seem, at least on most occasions, to be trying to get the listener in trouble. Still, it is a kind of encouragement, and that is the

 meaning-as-use of it.

(It should be acknowledged that mere encouragement is not

enough to enjoin the vocabulary of rights, which implies definite moral justification. For example, "You are my son, therefore I'll stand behind you, whatever you've done, right or wrong.")

Meaning-as-use therefore seems incompatible with any commitment to minds, times, numbers, or rights because it rejects reference to objects in favor of language as a practical activity of will. (indeed, for Schopenhauer, speech is a bodily action, a move.

Crucially: Wittgenstein's doctrine is also incompatible with

material, physical objects – forces, particles, energy packets. etc. (This is not to say that there is no sunshine, gravitational attraction, etc.!). Wittgenstein did not emphasize this, of course.

Will is not a 'thing,' either, not a mental object of reference. "Her heart is full of good will" might simply mean you can trust

her. There need be no ontological commitments here.

Schopenhauer was quite influential on Wittgenstein's view of language as an expression of will. Direct evidence of this influence is not easy to find; but see Wittgenstein, *Philosophical Grammar*

pp. 215-218, and the following remark from MS 158, p. 34v, in

The Big Typescript, p. 300e, in which Wittgenstein quotes Schopenhauer in English:

"Schopenhauer: *'If you find yourself stumped trying to convince someone of something and not getting anywhere, tell yourself that it's the will & not the intellect you're up against.'") p. 300e.*

Talking/writing is an activity of will, a movement of the body. Moreover, dogs bark, cow moo, roosters cock-a-doodle-doo – all expressing a will to live, a yearning to exist. I can feel that yearning within myself, too, as definitely as I can my own consciousness. It

is essential, it is what drives communication. Consciousness, Schopenhauer believed, is not all that's needed to explain the

world; there must also be motivation, action generated by will, otherwise no world could ever come to be. This seems quite true.

Wittgenstein specifies a multiplicity of ever more specific practical uses of words. There is nothing vague or vacuous about his list – giving orders, reporting events, speculating about events, and on

and on. It includes, between the lines, misreporting events, lies, promises, encouragement, hate speech, love poems, prayers, newspaper stories, books, ad infinitum.

So, again, although Wittgenstein's meaning-as-use undermines ontological claims such as "There are pains," "There are minds," "There are sense-data," and so on, it also, and just as importantly, shows that claims about material objects, physical objects, as

objects of reference, are suspect as well. Therefore, such claims as

"There are material objects," "Consciousness arises out of physical brain events," "The world is material and nothing more," must also be rejected. Wittgenstein did not stress this, and neither did the analytic philosophers, including Austin, Ryle, and Quine.

Recently, Alan Bustany said this about Gödel:

There are statements in Arithmetic that we know can neither be proven nor disproven.

*That is, we can **prove** things like 2 + 2 = 4, and we can*

***disprove** things like 2 x 2 = 5, but there are some expressions that we can't prove or disprove. I can't give you an example of such an expression, but a very clever man named Kurt Gödel **proved** that such expressions exist. Actual examples in Arithmetic are long and complicated.*

Mr. Gödel's genius was to find a way of representing "this statement cannot be proven" (let us call this statement S) as an arithmetic *expression by using a coding system that avoided referring to itself (because self-reference leads to nasty contradictions and paradoxes). We now call coding systems like*

*this a Gödel numbering. The system allowed him to prove that an arithmetic expression **equivalent to** S must exist.*

Informally, either S is true and we have a true statement that cannot be proven, or S is false and we have a contradiction

*(a false statement that we can prove). Then, assuming arithmetic does not have contradictions (it is **consistent**), we have a statement that cannot be proven or disproven.*

If you want to be a very clever ten-year-old, you will avoid saying

"There are true statements that cannot be proven" and instead say "There are statements that can neither be proved nor disproved."

That way you will avoid falling into some fallacious interpretations of Mr. Gödel's work that are rife on the Internet.

- From *Quora Digest*, Internet

My Conclusions:

Perhaps it is as simple as this: Arithmetic statements like

"2+3 = 5" are not about numbers, platonic universals, forms, etc. – mental objects – but are about adding two bricks to three bricks

and getting five bricks, or even occasions of thinking, changing one's opinions, etc. Worries about consistency, completeness, etc. are of no practical concern to most of us.

When we do algebra, we know that "x+2 = 5" means that when you add two items to x, whatever x represents (number of bricks or whatever), you get five items. That means x must = 3. That's a certainty.

Gödel believed in the existence of mental objects, numbers. Wittgenstein's meaning-as-use demolishes this view. Thanks to Gödel, Russell and Whitehead's *Principia Mathematica* was shown to be a hopeless project, which Wittgenstein already suspected. Hilbert's dream was dashed.

Applying meaning-as-use to Gödel's statements – in a

consistent way – shows that Gödel must have been thinking that statements are a kind of object, like bricks. That is false. We can

ask the weight of a pile of bricks, but what sense could it make to ask how much a statement weighs? The *sentences* used to express

a statement are merely what Russell would have called tokens of types. A sentence might have some physical weight, since it consists of marks on paper or sounds in air, but a *statement* has no particular size or weight. Add up all the *statements* held to be true about bricks and you will not get an answer in units of weight. [See Potter, Michael,

"*Wittgenstein on Mathematics,*" pp 122-137 in Kuusela, Oscari

and Marie McGinn, *The Oxford Handbook of Wittgenstein*, 2011, Oxford University Press].

What Gödel, Hilbert and Russell had to say is no longer of

much interest even to professional (academic) mathematicians. The project of putting mathematics on a "firm logical foundation" is

not crucial to the real use we put mathematics to – i.e. calculating, measuring, counting, determining trajectories, actuarial tables, profits and losses, interest rates, predicting the weather, etc. The

old philosophical concerns about consistency and completeness get little actual, practical mathematical work done.

So, Potter and Bangu are right about quite a lot, but not about

the "evident incoherence" of Wittgenstein's later philosophy of mathematics. At some point he realized he had bigger fish to fry

than Gödel and the misguided dream of Russell and Frege of

putting mathematics on a "firm logical foundation." He turned to

the real goal of his philosophy, certainty itself.

Wittgenstein, *On Certainty*, p. 39e: 309

"Is it that rule and empirical proposition merge into one another?"

Exactly. That is the solution to Potter's main question, and to Bangu's, as well!

What can we be certain about? Mathematics, of course, and that we are not now dreaming, and many other things. The concern with whether mathematical statements are tautologies or empirical claims falls away. His main concerns were finally captured as well as possible in *On Certainty*. I think he did not live long enough to gather all his philosophy up into a ball like his

Tractatus. But he knew he was on the right path, and that the rest of us might catch up later, unless we, too, run out of time.

BIBLIOGRAPHY

Bangu, Sorin, *Wittgenstein, Ludwig: Later Philosophy of Mathematics, Internet Encyclopedia of Philosophy*. http://www.iep.utm.edu/wittmath/

Gödel, Kurt, *On Formally Undecidable Propositions of Principia Mathematica And Related Systems*, Translated by B. Meltzer, Dover Publications, Inc. New York: 1992.

Kenny, Anthony, Editor, *The Wittgenstein Reader*, Blackwell Publishers, Cambridge: Mass.,1994.

Potter, Michael, "Wittgenstein on Mathematics," Chapter 6,

The Oxford Handbook of Wittgenstein, Edited by Oskari Kuusela and Marie McGinn, Oxford University Press, 2011.

Quine, W.V.O., *From a Logical Point of View*, Harvard University Press, Cambridge: Mass., 1953.

Wittgenstein, *On Certainty*, Edited by G.E.M. Anscombe &

G.H. von Wright, Basil Blackwell, 1969.

Wittgenstein, Ludwig, *Remarks on the Foundations Of Mathematics*, Edited by G.H. von Wright, R. Rhees, G.E.M. Anscombe, The M.I.T. Press, 1967.

Wittgenstein, Ludwig, *The Big Typescript: TS 213*, Edited and Translated by C. Grant Luckhardt and Maximilian A.E. Aue, Blackwell Publishing, 2005.

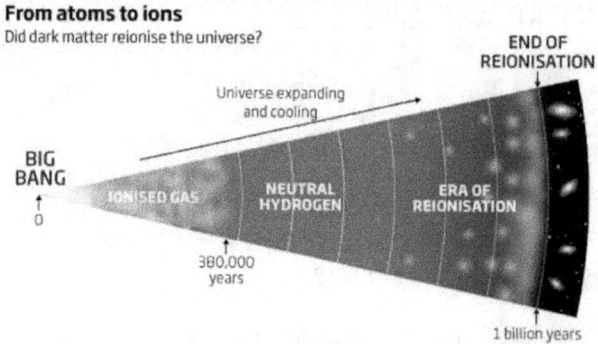

From atoms to ions
Did dark matter reionise the universe?

IV. The Big Bang Loves You

If we replace "God" with "the Big Bang" in Aquinas' Five Proofs, we get:

Proof One: The First Mover

We observe motion all around us. Whatever is in motion now was at rest until moved by something else, and that by something else, and so on. But if there were an infinite series of movers, all waiting to be moved by something else, then actual motion could never have got started, and there would be no motion. But there is motion now. So there must be a first mover which is itself unmoved. This First Mover we call the Big Bang.

Proof Two: The First Maker

Everything in the world has its efficient cause – its maker – and that maker has its maker, and so on. The coffee table was made by the carpenter, the carpenter by his/her parents, and on and on. But if there were just an infinite series of movers, all waiting to be moved by something else, then actual motion could never have got started, and there would be no motion now. But there is motion now. So there must have been a First Maker that was not itself made, and that First Maker we call the Big Bang.

Proof Three: The Proof from Necessary vs. Possible Being

Possible, or contingent, beings are those, such as cars and trees and you and I, whose existence is not necessary. For all such beings there is a time before they come to be when they are not yet, and a time after they cease to be when they are no more. If everything were merely possible, there would have been a time, long ago, when nothing had yet come to be. Nothing comes from nothing, so in that case there would be nothing now! But there is something now – the world and everything in it – so there must be at least one necessary being. This Necessary Being we call the Big Bang.

Proof Four: The Proof from Degrees of Perfection

We all evaluate things and people in terms of being more or less perfectly true, good, noble, and so on. We have certain standards of how things and people should be. But we would have no such standards unless there were some being that is perfect in every way, something that is the truest, noblest, and best. That Most Perfect Being we call the Big Bang.

Proof Five: The Proof from Design

As we look at the world around us, and ourselves, we see ample evidence of design – the bird's wing, designed for the purpose of flight; the human ear, designed for the purpose of hearing; the natural environment, designed to support life; and on and on. If there is design, there must be a designer. That Designer we call the Big Bang.

Copleston, F. S., *A History of Medieval Philosophy*, Harper & Row, New York, 1974.

1. Pegis, Anton C., (editor) *Introduction to St. Thomas Aquinas*, The Modern Library, New York, 1948.

2. Rather than deny the physical reality of the Big Bang, which I have done in earlier writings, I now simply equate it (It) with God and advocate that we think of the two interchangeably, as did Spinoza: *Deus sive Natura* (God or Nature).

Now I want to prove that God *exists* and *loves us*, and that we should trust Him to take care of us, and *align* our will with His, so that we need not fear death or suffering.

1. God exists – because God equals The Big Bang and The Big Bang exists, so

2. God loves us – because He *made* us

3. If God loves us – then we can trust Him.

4. If we can trust Him – then we need not fear death or suffering.

For those who enjoy symbolic logic, (the meanings of the letters should be obvious):

1. $B \equiv G$

2. B

∴ G.

3. M ⊃ L

∴ M

∴ L   (MP)

4. L ⊃ T

T ⊃ ~F

∴ ~F (HS)

⊃∴⊃

I also want to prove that we have Free Will, as against the claim (by Sam Harris and others) that Free Will is an illusion.

Sam Harris is concerned about the environment. He thinks we should *try* to have a clean environment. But if it makes sense to *try* to do something, it must make sense to *do* it (This is a point from Wittgenstein). For example, I cannot really *try* to rise and fly, because I *actually* cannot rise and fly. I cannot DO it!

1, T⊃D

2. T

∴D   (MP)

Maurice F. Stanley — Equations

V. A Terrible Sublimity: Kant and the Big Bang

Maurice F. Stanley

University of North Carolina at Wilmington

stanleym@uncw.edu

Abstract

I argue that Kant's claim, that concepts that apply only within the world of human experience cannot be legitimately applied beyond that realm, applies not only to judgments about God but also to the contemporary cosmologists' claims about the Big Bang. At best such concepts can serve as what Kant called "regulative ideas" that further the aim of achieving the most coherent empirical employment of reason.

Stephen Hawking argues that the origin of the physical universe is not off limits to science:

Many people would claim that the initial conditions of the universe are not part of physics but belong to metaphysics or religion. They would claim that nature had complete freedom to start the universe off any way it wanted ... . Yet all the evidence is that the universe evolves in a regular way according to certain laws. It would therefore seem reasonable to suppose that there are also laws governing the initial conditions. (quoted in Lightman, 1986)

Hawking, in a popularization of his research, describes these initial conditions:

At the Big Bang itself, the universe is thought to have zero size,

and so to have been infinitely hot. But as the universe expanded,

the temperature of the radiation decreased... (Hawking, 1988, p.

123)

Some two centuries ago Immanuel Kant offended metaphysicians and religious thinkers by arguing that certain ideas, among them

the idea of a beginning or first cause of the universe, are

transcendental, employing concepts outside the realm in which they apply, viz. the phenomenal world, the world of appearance:

The principle of causality has no meaning and no criterion for its application save only in the sensible world. (Kant, 1929, p. 511, A609 B637)

Hawking acknowledges Kant's view and devotes a paragraph to explaining and refuting it:

His [Kant's] argument for the thesis was that if the universe did

not have a beginning, there would be an infinite period of time before any event, which he considered absurd. The argument for

the antithesis was that if the universe had a beginning, there would be an infinite period of time before it, so why should the universe begin at any one particular time? ...They are both based on his unspoken assumption that time continues back forever, whether or not the universe had existed forever. As we shall see, the concept

of time has no meaning before the beginning of the universe. (Hawking, 1988, pp. 8, 9)

This is not exactly what Kant argued. His argument for the thesis that the world had a beginning is that otherwise an infinite series of tasks would have to be completed before the world could come to be, which would be impossible. And his argument for the antithesis is that if the world had a beginning, it would, as a beginning, have been preceded by an "empty time," a time when the world was not, But, Kant says, "No coming to be of a thing is possible in an empty time, because no part of such a time possesses, as compared with

any other, a distinguishing condition of existence rather than of nonexistence." I read this to mean what Aquinas meant by the maxim, *ex nihilo, nihil fit* -- "Out of nothing comes nothing." No

 part of such an empty time could give rise to something that exists. (If it could, that part of "empty time" would be special, distinct

from other parts, and therefore wouldn't be empty.)

So it does not seem to me that Kant's view depends upon an "unspoken assumption" that time extends back forever, though he surely thought that. His view depends on the reasonable belief that it is logically impossible to complete an infinite series of tasks, and that you can't get something from nothing. Kant's conclusion is that you can prove neither that there was a beginning to the world nor that there was not, because to speak of a "cause of the world" is to extend a concept that applies within the world of our experience, or possible experience, to the whole universe. Every event has a cause, but for Kant the whole universe is not just another event, one more object of our possible experience.

For Hawking, plainly, it is. And what's more he cites empirical evidence:

In 1929, Edwin Hubble made the landmark observation that wherever you look, distant galaxies are moving rapidly away from us, In other words, the universe is expanding. This means that at earlier times objects would have been closer together. In fact, it seemed that there was a time, about ten or twenty thousand million years ago, when they were all at exactly the same place, and when, therefore, the density of the universe was infinite. This discovery finally brought the question of the beginning of the universe into the realm of science, (Hawking, 1988, p. 9)

More putative evidence for the Big Bang has been discovered. On April 23, 1992, Dr, George Smoot, a Berkeley astrophysicist and leader of the research team associated with NASA's $4 million Cosmic Explorer satellite project, announced the detection of a gigantic, wispy cloud of microwave energy at the fringe of the observable universe, some 12 million light-years away, which he and the scientific community believe to have been produced by the Big Bang. Dr, Smoot was quoted in newspaper articles as saying, "What we have found is evidence for the birth of the universe." Subsequently newspaper editorials displayed a variety of reactions, most typically a mixture of awe and misunderstanding.

If these major discoveries really do constitute evidence of the Big Bang, Kant must have been mistaken in placing the beginning of the world outside any possible human experience. Kant believed that any claim about the beginning of the universe -- that there was one or that there was not one -- must involve an illegitimate application of a concept that applies only *within* human experience to the *whole* of experience or the universe itself. To assert that God created the universe in 4004 BC is unsupportable by any putative evidence because it applies the idea of cause (or creator, or maker) outside the realm where such concepts make sense. Everything was made or caused or created by somebody or something, we reason, so the whole universe was made by something, viz. God.

Hawking argues that everything in the world obeys physical laws, and so, he says, it is "reasonable to suppose" that the origin of the whole universe, the universe-as-awhole, obeys physical laws.

Maybe it does. But we can't know such a thing, any more than we could possibly know it was pink on the outside or hot on the inside, or such-and-such a size. Every physical object has a size, but it makes no sense to say the whole universe has a size or a

temperature because we can't compare its size or temperature with anything else. We can't hold a ruler or a thermometer up to it. To think otherwise is to commit the fallacy of composition, as

logicians call it.

But if we have, in those wispy clouds of radiation, real evidence of the Big Bang, all these considerations are wrong. Can we, as the scientific community now seems to believe, have empirical

evidence of an event which is itself in principle unobservable?

First we must establish that the Big Bang is indeed in principle unobservable. Though it will seem unfair to quibble with

Hawking's wonderfully sportsmanlike popularization of contemporary cosmology, still it is perhaps the best we can do as non-scientists who must depend on such popularizations. But it

may turn out that the popularization is misleading rather than that

the theory is wrong, or it may turn out that this theory is incapable

of being expressed in ordinary language.

When Hawking says the universe at Big Bang time had "zero size"

-- "a point of infinite density and infinite curvature of space-time" (1988, p. 140) at which all the laws of science would break down, also called a "singularity," it is hard not to feel puzzled. Could one ever see a zero-size, mathematical point? No, because such a point

 is essentially an abstraction, like Plato's forms, the ideas of justice, truth and beauty, and the definition of a circle. Something of zero size is not really a something, not a physical thing, at all. A lump

of sugar of zero size is not a real lump of sugar. It won't sweeten your tea. And how can such a "thing," of zero size, have a temperature? Could an idea, an abstraction, have a temperature?

Is the idea of Hell a hot idea?

Hawking admits that:

Because mathematics cannot really handle infinite numbers, ... the general theory of relativity predicts that at the singularity the theory itself breaks down.... This means that even if there were events before the Big Bang, one could not use them to determine what would happen afterward, because predictability would break down at the Big Bang.... As far as we are concerned, events before the Big Bang can have no consequences, ... we should therefore cut them out of the model and say that time had a beginning at the Big Bang. (1988, p. 50)

In this passage he entertains the possibility that there could be events before the Big Bang; but the theory can't predict them, *even if they existed,* nor would they have consequences (as far as we are concerned), so we must just forget them, calculate them away. For Hawking's theory, the nonexistence of events before the Big Bang has a kind of necessary truth.

To say there *must be* a sun is vastly different from saying there *is, in fact,* a sun. When scientists say there *must be* a Big Bang singularity, this is quite different from saying there *is* in fact, or *was* in fact, an actual Big Bang.

It is hard to avoid the suspicion that the Big Bang has a kind of theoretical necessity - as if there *must* be a Big Bang, as if there being no such thing would be unthinkable. Such certainty is the hallmark of analytic truth, systemic necessity, but not of empirical fact. In order that the Big Bang be a matter of fact, it would have to be possible that it not exist! Kant says that such necessity is, for human reason, a veritable abyss.

"Eternity itself, in all its terrible sublimity ... is far from making the same overwhelming impression on the mind...."

(1965, A613 B641)

Not only do the laws of physics break down at the Big Bang, but ordinary language does as well. The Big Bang is not really a "bang", for there was no atmosphere to carry sound waves; and how can it "expand" from no size to a larger size? How can it cool down from being infinitely hot? How can it be hot if there's no room for particles or protons to move (no conduction, no convection, no radiation)? It could have no parts to bump against one another. Wouldn't this no-size universe be infinitely cold? And if physical laws do not apply to it, why should it be bursting to expand as if Boyle's Law applied? And we might as well mention that "hot" and "curved" apply to real physical objects, not to abstractions or zero size points. Of course Hawking brings into the account quantum theory, which applies to very small stuff. But zero size is not "very small!"

Even if these concerns, or perhaps quibbles, can be cleared up by a more precise and technical choice of words, the following problem indicates that there is a more serious difficulty with the concept of the Big Bang, at least if it is to be understood as Hawking seems to understand it. If the notion scientists have, that looking into space is looking backward in time, is true, might we in principle build a telescope that would allow us to see 15 billion light-years out into space? We could see the Big Bang itself (well, not if it has, or had, zero size, but perhaps at least as it was on the tail-end of the first moment). Now that would be the whole universe at that moment, which contained everything that ever was and ever would be -- scrunched up a bit, of course, but including us (or at least our physical "constituents"). So we would be looking back at a former state of ourselves, which is surely an incoherent notion, as if we might in principle be able to watch ourselves being born, say.

We resulted from the Big Bang, or at least the stuff we're made of, did. But this, this world as it is now, *is* the Big Bang, still expanding; and here we are, 15,000,000,000 light years away. So we're here on earth and also there, across the universe. Such concepts as time and space lose their ordinary use in this mad tea party. I can't look at myself sitting on the steps of a house a mile away, even through the fanciest new telescope! I would have to be here and also a mile away from myself. One way out of this problem would be to redefine "myself" as a space-time line that includes my whole past and future, etc., but then we've left ordinary language far behind, again; and besides that we have begged the philosophical question of the nature of time (a fixed continuum or a flowing river?)

Imagine a botanist who observes a certain tree for awhile and concludes that it is growing, and then theorizes that it must have been smaller in the past. She concludes that at some long-ago point in time -- indeed at some definite time -- the tree must have been a point of no size at all, but heavy, infinitely heavy, etc.

We know about acorns, of course, so this is not a persuasive story. The Big Bang is not an acorn and the universe is not a tree, etc. But I submit that there is a certain weak but suggestive parallel, at least if we think of the Big Bang as essentially unobservable and therefore purely theoretical.

A police officer sees a circular patch of light on a wall beside her and infers that the source of the light must be a burning point of no size at a certain point beyond her, etc. Of course she then realizes the source of the light is a flashlight.

Scientists believe, however, that the existence of those wispy microwave clouds is evidence for the

Big Bang. But how can you find evidence of something which is *in principle* unobservable?

In an article for *US News* Gregg Easterbrook discusses recent accretions to the Big Bang theory. Andrei Linde, for example, a physicist at Stanford University, offers a cosmos that can have no end because it copies itself endlessly. That optimistic prognosis, Easterbrook cautions us, is all conjecture. How, indeed, could it be anything *else? Easterbrook says:*

Much of the mathematics of the big bang is based on hypothesis. Assumptions about unobservable regions are subjected to pages of number-crunching, creating the quasi result in which fuzzy notions and highly precise formulations are combined, leaving everyone unsure about whether the final product owes more to precision or guesswork. And the ever changing findings of cosmology make genesis theories moving targets.[MIT physicist Alan] Guth has been trying to incorporate into his theory the newly emerging evidence, mainly from studies of distant supernovas, that the cosmic expansion is actually speeding up, rather than slowing down, as most theorists expected. Saul Perlmutter, an astrophysicist at Lawrence Berkeley National Laboratory, says, "Big-bang theory tends to go overboard for whatever the latest new idea is, whereas experimenters like me tend to assume that the universe is extremely complicated and each new item of information we get only shows us how much we don't know."

One thing we don't know is why there is a cosmos at all. As Derek Parfit, a fellow at

Oxford University, has written, "No question is more sublime than why there is a universe: Why is there anything, rather than nothing?"[actually Heidegger's question! M.S.] Just try to conceptualize true nothingness: that there had never been anything. Probably there always had to be something, because the absence of existence is not possible; the question is how far back one must go to locate the ultimate antecessor. (Easterbrook, p. 22)

Evidence, Kant would say, is a concept that can operate only within the world of our experience, to which our scientific concepts apply. Such concepts, he says, can be employed ... to explain the possibility of things in the world of sense, but not to explain the possibility of the universe itself. Such a ground of explanation would have to be outside the world and could not therefore be an object of a possible experience. None the less, though I cannot assume such an inconceivable being as existing in itself, I may yet assume it as the object of a mere idea, relatively to the world of sense. (1965, A677 B705)

Such concepts Kant calls "regulative ideas."

We might observe the beginning of a star, but not the beginning of the universe, of the world thought of as a totality. This is because the idea of the universe is a regulative idea. Kant offers three examples of such regulative ideas: the "I" itself, the world as a totality, and God. These ideas lie outside all human experience, but at the same time they help direct our inquiries. They are heuristic.

The Big Bang is such an idea, the idea of the universe thought of as a totality. Our empirical ideas apply to the series of causes (conditions) leading up to the totality, but not the totality itself. This series of conditions can be regarded "as if it had an absolute beginning [the Big Bang], through an intelligible cause. All this shows that the cosmological ideas are nothing but simply regulative principles ... ." (1929, A670686, B698-714)

Regulative ideas have been considered by philosophers since Kant as maxims or rules of thought which come out of our fundamental purposes of inquiry – as in, "always look for the cause." Regulative principles guide our inquiries even though we have no proof that they are true. Kant believed it was rational to look and hope for a complete coherent system of thought, although we have no *a priori* guarantee that it could be found.

In roughly the same way the "I" which can never be observed is a regulative idea that leads us to the rules of morality, and the idea of God is a regulative idea that gives us hope that justice will finally prevail in the world. Scientists discover laws, but do not like to see them in disarray, unrelated to one another. They want them to come together, one law under another, in some order.

Such concepts as reality, causality, etc., Kant says, "even that of necessity in existence, apart from their use in making possible the empirical knowledge of an object, can be employed ... to explain the possibility of things in the world of sense, but not to explain the possibility of the *universe itself*. Such a ground of explanation would have to be outside the world, and could not therefore be an object of a possible experience." (1965, A677-A678; B705-B706)

Such regulative principles guide scientific inquiry even though we cannot know whether or not they are actually true. Naturally scientists are always looking for a coherent system of knowledge and Kant believed this is rational, even though we cannot allow ourselves to think that such a regulative principle, in this case the Big Bang (the universe thought of as a totality), can be anything more than a guide to inquiry.

These regulative principles may be useful or even necessary to ensure the greatest, most coherent empirical employment of reason, but they must not be taken as standing for anything actual or real. The idea of God as a supreme, purposive intelligence can guide our reason in its search for teleological connections in nature, but to try to describe such a being by means of empirical concepts is to engage in counterproductive self-delusion. The Big Bang is rather like God in this regard. It is an idea that directs our thought, but it can never be observed. As Kant puts it:

If ... we ask first, whether there is anything distinct from the world, which contains the ground of the order of the world and of its connection in accordance with universal laws, the answer is that there undoubtedly is. If, secondly, the question be, whether this being is substance, of the greatest reality, necessary, etc., we reply that this question is entirely without meaning.... If, thirdly, the question be, whether we may not at least think this being ... in analogy with the objects of experience, the answer is certainly, but only as object in idea and not in reality. (1865, A697 B725)

I conclude, as I believe Kant would, that the idea of the Big Bang is no more than a regulative idea taken too far, an attempt to use scientific concepts that can only apply within the universe -- i.e. within the world of possible human experience -- to the totality of that universe. As Kant recognized, we cannot help making such attempts, but when we do we land in a soup of nonsense. And scientific nonsense is no better (and no worse) than religious or metaphysical nonsense.

References

Easterbrook, Gregg. "Before the Big Bang." In *Mysteries of Outer Space. US News & World Report,* 2003, p. 22.

Hawking, Stephen W. *A Brief History of Time.* London, Bantam Press, 1988.

Kant, Immanuel. *Critique of Pure Reason.* Translated by Norman Kemp Smith. New York, St Martin's Press, 1929.

Lightman, Alan. "The Origin of the Universe." In Timothy Ferris, editor, *The World Treasury of Physics, Astronomy, and Mathematics.* Boston, Little, Brown and Company, 1991.

Maurice F. Stanley   Equations

End of EQUATIONS

www.ingramcontent.com/pod-product-compliance
Lightning Source LLC
Chambersburg PA
CBHW070655220526
45466CB00001B/455